# Never Stop Wondering

BY EMILY MORGAN

illustrated by Jacqui Crocetta

NSTA Kids
National Science Teachers Association
Arlington, Virginia

Claire Reinburg, Director
Rachel Ledbetter, Managing Editor
Andrea Silen, Associate Editor
Jennifer Thompson, Associate Editor
Donna Yudkin, Book Acquisitions Manager

**ART AND DESIGN**
Will Thomas Jr., Director

Illustrated by Jacqui Crocetta

**PRINTING AND PRODUCTION**
Catherine Lorrain, Director

**NATIONAL SCIENCE TEACHERS ASSOCIATION**
David L. Evans, Executive Director

1840 Wilson Blvd., Arlington, VA 22201
*www.nsta.org/store*
For customer service inquiries, please call 800-277-5300.

Copyright © 2019 by the National Science Teachers Association.
All rights reserved. Printed in Canada.
22 21 20 19      6 5 4 3

Lexile® measure: 790L

*NSTA is committed to publishing material that promotes the best in inquiry-based science education. However, conditions of actual use may vary, and the safety procedures and practices described in this book are intended to serve only as a guide. Additional precautionary measures may be required. NSTA and the authors do not warrant or represent that the procedures and practices in this book meet any safety code or standard of federal, state, or local regulations. NSTA and the authors disclaim any liability for personal injury or damage to property arising out of or relating to the use of this book, including any of the recommendations, instructions, or materials contained therein.*

**PERMISSIONS**
Book purchasers may photocopy, print, or e-mail up to five copies of an NSTA book chapter for personal use only; this does not include display or promotional use. Elementary, middle, and high school teachers may reproduce forms, sample documents, and single NSTA book chapters needed for classroom or noncommercial, professional-development use only. E-book buyers may download files to multiple personal devices but are prohibited from posting the files to third-party servers or websites, or from passing files to non-buyers. For additional permission to photocopy or use material electronically from this NSTA Press book, please contact the Copyright Clearance Center (CCC) (*www.copyright.com*; 978-750-8400). Please access *www.nsta.org/permissions* for further information about NSTA's rights and permissions policies.

**Library of Congress Cataloging-in-Publication Data**

Names: Morgan, Emily R. (Emily Rachel), 1973- author.
Title: Never stop wondering / by Emily Morgan.
Description: Arlington, VA : National Science Teachers Association, [2019]
Identifiers: LCCN 2018011440 (print) | LCCN 2018017317 (ebook) | ISBN 9781681403106 (e-book) | ISBN 9781681400082 (print) | ISBN 9781681406503 (library)
Subjects: LCSH: Science--Methodology--Juvenile literature. | Nature observation--Juvenile literature.
Classification: LCC Q175.2 (ebook) | LCC Q175.2 .M67 2018 (print) | DDC 507.2/1--dc23
LC record available at *https://lccn.loc.gov/2018011440*

# Dedication

*To my parents, Vonda and Jim Stevens, who approach the world with open minds and open hearts.*
**—Emily Morgan**

*For my mother, Nancy Crocetta, who taught me to be curious.*
**—Jacqui Crocetta**

*I would rather have questions that can't be answered than answers that can't be questioned.*
**—Richard Feynman**

Never stop wondering, never stop questioning.
Never stop trying to figure things out.

Always keep searching, always keep asking.
That's what science is all about.

Why do rainbows form?

What makes a plane fly?

How do magnets work?

When one is young, the questions come fast: the how, why, who, what, and when.

But as some people grow, the questions don't flow in their minds like they did way back then.

When some turn into adults, the aim of their thoughts is no longer to learn how, why, or who.

They become busy with lots and the wondering stops. Don't let that happen to *you!*

The great thinkers through time kept their wondering minds fixed on questions of cause and effect.

Some never outgrow the desire to know how all of the pieces connect.

Sir Isaac Newton
Physicist and Mathematician

So study the world, look for changes and patterns, and try to predict what comes next.

Ask why things happen, then test your ideas, and persist with the ones that perplex.

Notice the structure and function of objects in nature and human design.

Calculate scale, proportion, and quantity, like length, size, distance, and time.

Often you'll have more questions than answers, and it's OK to say, "I don't know."

Take time to delight in and savor the wonder and think of the next route you'll go.

If you find out you're wrong, don't get discouraged, and whatever you do, don't give up.

Celebrate what you learned, try another idea, keep your eyes, ears, and mind opened up.

Each day brings new questions and new explorations. Each day, brand new facts are discovered.

Ideas begin forming, models take shape, and they change as more evidence is uncovered.

With some questions it's easy to look up the answers, 'cause someone once asked and found out.

Other questions you might be the first one to think of, the first one to wonder about!

You'll see that the most compelling of questions keep your mind occupied.

But it may be a surprise to also find out that your heart and soul are along for the ride.

You see, science isn't dull, boring, or dry. It's exciting, fun, and inspiring.

There's delight in the mystery, satisfaction in knowing, and joy with each new discovery.

Over the years, the great questions of science have fueled the strongest of passions—

from discovering pi to learning to fly to inventing the newest contraptions.

We'll never learn all there is to know about how this world of ours works.

But each new discovery is a piece of the puzzle, from planets to atoms to quarks.

21

Just because someone says it or writes it doesn't always mean that it's true.

Before you buy in, before you accept it, look for the facts that prove it to you.

It's OK to doubt as
you figure it out.
Don't instantly believe
something's true.

Go to the sources you know you can trust,
the ones that have been reviewed.

You might meet some folks who "have all the answers," but that's just an act, just for show.

The key to science and true understanding is realizing how much we *don't* know.

Lightning never strikes the same place twice. MYTH

MYTH

People get warts from frogs and toads. MYTH

Humans can't grow new brain cells. MYTH

The more you learn, the more questions you'll have ... and that's just all part of the fun.

Learn all you can, and soon you'll discover the work of science will never be done.

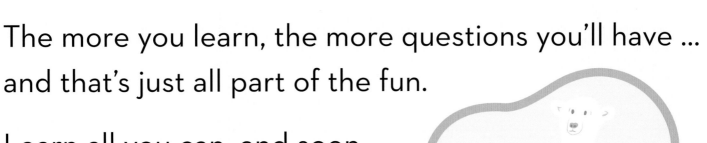

MYTH — Climate change isn't happening.

MYTH — Bats can't see.

MYTH — It takes seven years to digest gum if you swallow it.

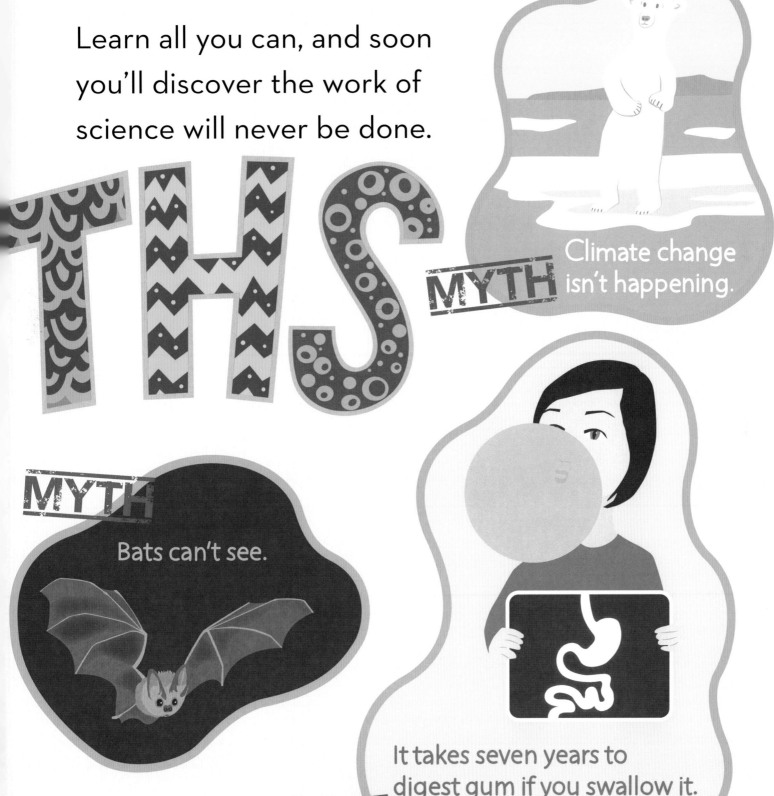

Life will always give us questions to answer and problems we have to resolve.

And the quest will continue as the answers we find lead to more mysteries to solve.

Around 80% of the ocean remains unexplored.

# Wondering

Just remember, keep wondering; remember, keep questioning; remember, keep trying to figure things out.

What do *you* wonder? What do *you* question? What do *you* want to find out?

WHY?

HOW?

# Ideas to Encourage Children to Wonder

### Experience Before Questioning

During my 20 years in education, I encountered one strategy that I find essential for encouraging children to ask good questions: **giving them an experience with a topic first.** It is hard to ask questions about something you have no experience with or have not experienced in a long time. However, giving students time to explore and tinker with objects related to the topic can work wonders. Simply opening a lesson with the question "What are you wondering about magnets?" is not likely to produce many good wonderings. However, if you launch a lesson by giving students time to explore with magnets, the questions will flow. This hands-on approach might not work for certain topics, such as the solar system or ocean ecosystems. For these subjects, showing students a thought-provoking video clip or an engaging app could elicit the same response.

### O-W-L Chart

An **O-W-L chart** is like a K-W-L chart, with one big difference—instead of writing what they know about the topic in the first column, students record what they observe. In the second column, they record what they're wondering. In the third column, they write what they have learned. Karen Ansberry and I introduced the O-W-L chart in our book *Picture-Perfect Science Lessons* (NSTA Press, 2010) and have had great experiences with it over the years. We encourage students to use all of their senses (except taste!) when making observations for the O column and to make both qualitative and quantitative observations.

To me, the best part of using an O-W-L chart is that students realize as they fill in the *L* column that they have more questions for the *W* column ... and the learning can go on and on! O-W-L charts can be used with objects like owl pellets, Mexican jumping beans, wind-up toys, and magnets, or with natural phenomena like sunsets or Moon phases.

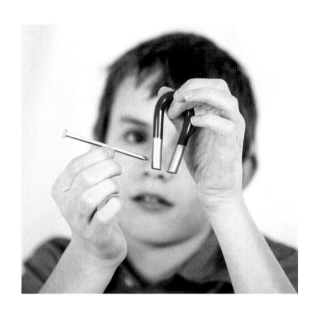

## Parking Lot

Do you ever have a student who asks a question in the middle of a lesson or read-aloud that has *nothing* to do with the topic? But it is a really good question! A **parking lot** is the perfect place for this kid to put their question. Have the student write the question on a sticky note and place it on a poster titled "Parking Lot." This way you can validate the question, celebrate it by displaying it for everyone to see, and revisit it later when time permits.

## Wonder Tree

A **wonder tree** is a great way to capture initial questions about a new topic and see how questions continue to emerge as you learn more. Create a wonder tree by cutting out a tree pattern from craft or poster paper and cutting out leaf-shaped pieces of paper where students can record their questions. On the first day of a new unit, give students an initial experience with the topic. This can be a hands-on or digital experience—just make sure it is interesting! Then, have students record their questions on the leaves and tape them to the tree. (This could be used along with the O-W-L chart; students could choose their most compelling question to display on the tree.) Encourage new questions to be added throughout the unit and you will all see the wonder tree "grow" as you learn. This is a great visual representation of the idea that learning and wondering go hand in hand—the more you learn, the more questions you have.

## Question Box

When I was in college (before Google!), my Biology 101 professor introduced a **question box** during her first lecture. She invited us to place questions in the box before or after class, and then she would select one or two to answer prior to the following lecture. The only rule was that the questions had to be about biology. She made overheads (remember those?) of the questions so we could see them in the students' handwriting. She read the question and then answered it. Most of these questions had nothing to do with the current topic we were studying. I can only imagine the questions she read through while trying to select one. One time she chose my question: "Why does eating ice cream too fast give you a headache?" (I was asking for a friend.) This five-minute period at the beginning of her lectures was like magic. I never wanted to be late, never wanted to miss her answers, and wanted to see if my question might have been selected out of the one hundred or so she had read through.

When visiting a school in Dallas, Texas, a few years ago, I noticed that the art teacher, Talitha Kiwiet, had a light-up question mark on her bulletin board. When students added a new question, they got to turn the light on. She said this practice helps inform her of students' interests and often leads to new explorations. Kids were excited when they saw that light go on. I thought it would be a great addition to the question box!

Photo credits: Shutterstock (magnet); Author (parking lot, question box); Jen Molitor (wonder tree)

## Question Sort

So you've got a great question ... now what? Well, it depends on what type of question it is! One way to help students understand that different types of questions require different actions is a **question sort.** Karen Ansberry and I featured the question sort in our book *More Picture-Perfect Science Lessons* (NSTA Press, 2007).

Here's how it works: Give students a common experience with a topic using thought-provoking objects, readings, video clips, etc. Ask students to write questions they have about the topic on sentence strips or sticky notes large enough for everyone to read. Explain that the way a scientist goes about answering a question depends on the kind of question he or she asks and that one way to sort questions is by dividing them into two categories: research questions and testable questions.

**Research questions** can be answered by using resources of scientific information such as books, reliable websites, or experts.

**Testable questions** can be answered by observing, measuring, or doing an experiment.

Create a T-chart with the columns labeled "Research Questions" and "Testable Questions." Read each question aloud and, as a group, decide which category it goes in. Then, tape the question to the chart in the appropriate category. After sorting the questions, you can choose one for students to investigate as a class in a guided way or have teams choose a question to explore on their own. This might depend on time and the supplies you have on hand. Check the website below for more examples of question sorts and ideas for choosing and investigating some of the questions.

For more detailed descriptions of these activities and printable student pages, go to:
*www.nsta.org/neverstopwondering.*

> *The pursuit of truth and beauty is a sphere of activity in which we are permitted to remain children all our lives.*
> **—Albert Einstein**